FARM ANIMALS
GOAT

Katie Dicker

A⁺
Smart Apple Media

Published by Smart Apple Media,
an imprint of Black Rabbit Books
P.O. Box 3263, Mankato, Minnesota, 56002
www.blackrabbitbooks.com

Printed in the United States of America,
at Corporate Graphics in North Mankato, Minnesota.

Designed by Helen James
Edited by Mary-Jane Wilkins

Library of Congress Cataloging-in-Publication Data

Dicker, Katie.
 Goat / Katie Dicker.
    p. cm. -- (Farm animals)
 Includes bibliographical references and index.
 ISBN 978-1-62588-021-5
1. Goats--Juvenile literature. I. Title.
 SF383.35.D53 2014
 636.3'9--dc23
                       2013000059

Photo acknowledgements
l = left, r = right, t = top, b = bottom
title page Eric Isselee/Shutterstock; page 3 chochi_cz/Shutterstock;
4, 5, 6 iStockphoto/Thinkstock; 7 Olga Selyutina/Shutterstock;
8 iStockphoto/Thinkstock, 9 Ignite Lab/Shutterstock; 10 Svetlana Yudina/
Shutterstock; 11 iStockphoto/Thinkstock, 12 Hemera/Thinkstock;
13 TOMO/Shutterstock; 14 Aubrey Laughlin/Shutterstock; 15t kool99,
b WilleeCole /both Shutterstock; 16 Aggie 11/Shutterstock;
17 iStockphoto/Thinkstock; 18 James Michael Dorsey/Shutterstock;
19t iStockphoto/Thinkstock, b Lilyana Vynogradova/Shutterstock;
20t Kydroon, b Emily B/both Shutterstock; 21t Martin Lehmann,
b Teerinvata/both Shutterstock, r Stockbyte/Thinkstock; 22 IrinaK/
Shutterstock; 23 Four Oaks/Shutterstock
Cover Nataliia Antonova/Shutterstock

DAD0507
052013
9 8 7 6 5 4 3 2 1

Contents

My World

Bleat!

I am a goat.
I live on a farm
with lots of
other goats.

Here are some of my herd.

Goats do not like to live alone.
We stick together in groups.

Different Places

We can live in hot and cold parts
of the world. Our hair keeps us
warm in winter and cool in summer.

Goats usually
have a coat
of white, gray,
red, brown,
or black hair.

We are good at climbing and keeping our balance!

Our strong hooves help us to walk on steep mountains, rocky surfaces, or wet ground.

Cool Coats

When the weather is warm, the farmer cuts our coats. Our hair can be used to make fabric.

The soft, fine hairs of these Cashmere goats are combed or plucked to remove them.

The wool from these Angora goats is called mohair. It is soft, strong, and shiny.

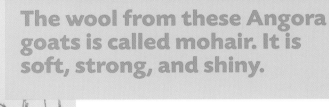

Our hair is cleaned, colored, and spun into thread. Then it is made into clothes and carpets.

Time to Eat

I like to eat leaves, twigs, grass, and grain.
I graze for most of the day.

Munch!

Goats have four parts to their stomach to help them digest the plants they eat.

We need plenty of fresh water, too. Sometimes we drink from running streams.

Baby Goats

Baby goats, called kids, are born in spring when the weather is warm and there is lots of grass to eat.

Slurp

Young goats drink their mother's milk. Some kids are bottle-fed.

Female goats often have twins. The farmer makes sure the kids are born safely.

A male goat is a buck (or billy), and a female goat is a doe (or nanny).

13

Who Looks After Us?

Our shelter keeps us safe at night. Inside, we stay warm and dry in winter weather.

The farmer cleans our shelter and gives us fresh straw to lie on.

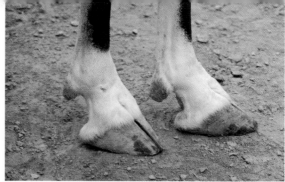

A vet visits regularly to keep us healthy and trim our hooves.

15

The Milking Parlor

We go to the milking parlor twice a day. The farmer puts milking machines on our udders.

Our milk can be used to make butter, cheese, yogurt, and cream.

Goats can make up to two gallons of milk every day.

On some small farms, goats are still milked by hand.

Farm Produce

Goats are farmed for their meat, milk, wool, and leather.

When goats reach a healthy weight, the farmer takes them to market.

People who are allergic to cow milk sometimes drink goat milk instead.

Goat cheese is very popular in France, where it is called chèvre.

Goats Around the World

Blackneck, Switzerland

Anglo-Nubian, England

Farmers in countries all over the world keep goats. Some goats live in the wild and others are kept as pets. Here are some of the different breeds.

20

Alpine, France

Rocky Mountain, USA

There are about 600 different types of goats around the world.

Golden Guernsey, England

21

Did You Know?

Goats are playful creatures and can be easily trained.

Male goats are very smelly! Their scent is strongest during the breeding season.

Goats usually live for about 12 to 15 years.

A goat has no top teeth at the front of its mouth.

Useful Words

grain
A type of cereal used for animal feed.

graze
Animals graze when they eat grass.

herd
A group of animals that live together.

leather
A material made from the skin of an animal.

vet
A doctor who takes care of animals.

Index

Web Links

www.animalcorner.co.uk/farm/goats/goat_about.html
www.kidskonnect.com/subjectindex/13-categories/animals/35-goats.html
www.agriculture-4-u.co.uk/pages/Livestock/goats/goats.php
www.kidcyber.com.au/topics/farmgoat.htm
www.kidsfarm.com/goats1.htm